Housing 70

快乐而健康
Happy and Healthy

Gunter Pauli

冈特·鲍利 著

唐继荣 译

丛书编委会

主　任：贾　峰
副主任：何家振　郑立明
委　员：牛玲娟　李原原　李曙东　吴建民　彭　勇
　　　　冯　缨　靳增江

丛书出版委员会

主　任：段学俭
副主任：匡志强　张　蓉
成　员：叶　刚　李晓梅　魏　来　徐雅清　田振军
　　　　蔡雩奇

特别感谢以下热心人士对译稿润色工作的支持：

姜竹青　韩　笑　杨　爽　周依奇　于　哲　阳平坚
李雪红　汪　楠　单　威　查振旺　李海红　姚爱静
朱　国　彭　江　于洪英　隋淑光　严　岷

目录

快乐而健康	4
你知道吗？	22
想一想	26
自己动手！	27
学科知识	28
情感智慧	29
艺术	29
思维拓展	30
动手能力	30
故事灵感来自	31

Contents

Happy and Healthy	4
Did you know?	22
Think about it	26
Do it yourself!	27
Academic Knowledge	28
Emotional Intelligence	29
The Arts	29
Systems: Making the Connections	30
Capacity to Implement	30
This fable is inspired by	31

城市里的水脏了,谁也不敢喝这种水。水上飘(水黾)拜访了青蛙一家,讨论发生的事情。

"一开始,人类只是把洗涤剂排入水中,但是现在情况更糟糕了,他们甚至把药品和激素也扔进我们珍贵的饮用水里。"水上飘叹息道。

The water in the city is dirty. No one dares to drink it. The water striders visit the frog family to discuss the situation.

"First, humans only threw soap in the water, but now it's worse – they're even throwing medicine and hormones into our precious drinking water!" laments the water strider.

城市里的水脏了

The water in the city is dirty

他们知道自己在干什么吗……

Do they know what they're doing...

"他们知道自己在干什么吗?"青蛙很好奇。

"不!他们没有意识到即使吃下肚,药丸仍然有效。更糟的是,过期的药丸被冲进厕所下水道,以致我们也被迫吃下它们!"

"那可不行。对水进行清洁处理代价高昂,而且药丸本身也要花很多钱呢。"

"Do they know what they're doing?" wonders the frog.

"No, they don't realise that their pills continue to work even after they've taken them. Worse, when pills are out of date, people flush them down the toilet so that we have to take them!"

"It doesn't make sense. It's expensive to clean water, and pills cost a lot of money."

"对水进行清洁？没有哪种水处理设备和化学手段能强大到中和水里的人造化学物质。"水上飘难过地说。

"这真是坏消息！"

"那么，人们为什么不用更天然的药物？阿育吠陀医药已经流传了几千年，而且这种药能像食物那样溶解。"

"Clean water? No water treatment or chemical is strong enough to neutralise manmade chemicals in water," the water strider says sadly.

"That's bad news!"

"So why aren't people using more natural medicine? Ayurvedic medicine has been around for thousands of years, and it dissolves like food."

阿育吠陀医药

Ayurvedic medicine

……中国有一套独特的健康传统……

...Chinese have a unique health tradition...

"阿育……什么？就是那种来自印度的植物药吗？"青蛙问道。

"阿育吠陀。没错，就是它。"

"我希望这种伟大的药物有个我能记住的名字。但我的确知道中国有一套强身健体的独特健康传统，让人不生病。"

"Ayu– what? Is that the plant medicine from India?" asks the frog.

"A-yu-r-ve-dic. Yes, it is."

"I wish this great medicine had a name I could remember. But I do know that the Chinese have a unique health tradition of strengthening the body so that it doesn't get sick."

"中国人有伟大的智慧,非洲人、阿拉伯人以及许多其他著名或不著名的文化也是。"水上飘沉思道。

"问题在于西方对天然药物持怀疑态度!"

"嗯,西方医学也有许多改善生活质量的发明。"

"The Chinese have a great wisdom, so do Africans, Arabs and many known and unknown cultures," the water strider muses.

"The problem is that the West is suspicious of natural medicine!"

"Well, Western medicine also has many innovations that improve quality of life."

对天然药物持怀疑态度

Suspicious of natural medicine

多少物种灭绝了

How many species go extinct

"这些发明没有改善我的生活质量。"青蛙说道,"你有没有意识到由于他们这种生活方式,有多少物种灭绝了?"

"你可以关注坏的方面,但你也可以看到好的方面。要知道,没有哪件事情是完美的,任何事物都可以提高。"

"They haven't improved my quality of life," says the frog. "Do you realise how many species go extinct because of their way of life?"

"You can focus on the bad, but you can also look for the good, knowing that nothing is perfect, and everything can be improved."

"那是不是意味着没有哪种药物能对所有人在所有时间都有效？"

"首先，我们要记住，许多人生病是因为他们有压力，又生气。"

"你的意思是，生病并不总是因为细菌或病毒引起的？"

"Does that mean there's no medicine that works for everyone, all the time?"

"Firstly, we need to remember that many people get sick because they are stressed and angry."

"You mean it's not always because of bacteria or viruses?"

因为压力和怒气而生病

Sick from stress and anger

乐观和快乐

Be positive and happy

"当然不是!其次,某些人觉得病了,只是他们认为自己生病了而已。"

"天啊!你是说这些人想象某件事情要发生,然后这件事情就真的发生了?"

"第三,当人生病时,如果相信自己会好起来,他们恢复健康就会更快!"

"这么说,最好的药物就是乐观和快乐,即便身边有坏消息也是如此?"青蛙问道。

"So! And secondly, some people feel sick because they think they're sick."

"Oh dear! You mean they imagine something that then comes true?"

"Thirdly, when people are sick, they get healthy faster if they think they can get better!"

"So the best medicine is to be positive and be happy, even when there's bad news around you?" asks the frog.

"噢，如果听新闻，我保证你所听到的一切都是坏消息。请记住，在中国，危机意味着机遇！"

"有能让人快乐的药吗，水上飘？"

"可能有！你可以从讲笑话开始每一天，经常笑，尤其是自己犯错时更要笑。还要一直想着，今天能为其他人做些什么！"

……这仅仅是开始！……

"Oh, I guarantee that if you listen to the news, bad news is all you hear. Remember, in China crisis means opportunity."

"Is there a medicine to be happy, Water Strider?"

"Maybe. You can start the day by telling a joke, laugh regularly – especially about your own mistakes – and always wonder what good you can do for others today!"

... AND IT ONLY HAS JUST BEGUN!...

……这仅仅是开始！……

...AND IT HAS ONLY JUST BEGUN!...

Did You Know?

你知道吗？

When people take pills, part of the medication passes through their bodies and is flushed down the toilet. Water-treatment plants cannot remove all traces of this medication.

当人们吃药时，一些药物成分通过身体后，被冲入厕所。水处理设备不能去除所有残留的药物成分。

Pharmaceuticals in water cause damage to people and wildlife. Drinking water can contain an accumulated cocktail of drugs.

水中的残留药物对人类和野生生物造成损害，而饮用水可能含有各种各样积累的药物，成了"药品鸡尾酒"。

Drugs are active beyond their expiration date.

过了有效期之后，药品仍具有活性。

Soap in water reduces water tension. This means that water becomes "wetter". And very wet water eliminates the natural protection we all have, while it permits water to penetrate fibers and take away the dirt.

水里的肥皂降低了水的张力。这就意味着水变得"更湿"了。非常湿的水消除了原有的自然保护，让水渗透到纤维中，把污垢带走。

Ayurvedic medicine originated in India 3 000 years ago. The term means "The science of life".

3000 年前，阿育吠陀医药起源于印度，这个名词的原意是"生命的科学"。

People who are happy and have a positive outlook on life are less likely to get sick.

那些快乐和有积极人生观的人不容易生病。

In China, a crisis is also considered an opportunity. The key is to have the capacity to see the opportunity in a crisis.

在中国，危机也被看作机遇，关键是要具备从危机中看到机遇的能力。

If all else fails, we should at least have the capacity to learn a lesson from that failure.

如果一切都失败了，我们至少应该有能力从失败中汲取教训。

Think About It
想一想

When you are sick in bed, do you feel happy?

当你卧病在床的时候，你感到快乐吗？

当你面临无法解决的麻烦时，你会有压力吗？

Do you get stressed when you have a problem you cannot resolve?

When everything goes wrong, do you spare a moment to think about the lessons you could learn, or do you prefer to forget the situation as soon as possible?

当事事不顺心时，你是会抽空思考能够汲取的教训，还是宁可尽快忘记这一困境？

一个国家能通过计算它的国民有多么幸福来衡量它的进步吗？

Can a nation measure progress by calculating how happy its citizens are?

Check the medicines you have at home. Does your family only take prescription drugs, or do some members of your family use alternative medicines? Check that your family's medicine is stored in a safe place, and then separate the bottles into two groups: those prescribed by a medical doctor, and those bought over the counter. Look at the expiration date of the drugs. Are some too old to be taken? The medicine that is too old needs to be disposed of. How would you do that?

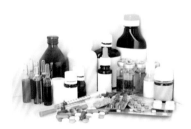

检查家里的药品。你的家人只服用处方药,还是有些人会服用非传统药品?检查并确保你家的药物放在一个安全的地方,然后将药瓶分为两组,一组是医生开的药,一组是在药店买的药。注意药物保质期,有些是不是已经过期了?过期的药需要正确处置。那么,你是怎样处置它们的?

TEACHER AND PARENT GUIDE

学科知识
Academic Knowledge

生物学	由腺体分泌的激素调节生理和行为；只要水的张力维持正常，水黾（俗名水上飘）就能在水面上行走；青蛙的皮肤有一层天然的保护膜，只要表面张力正常，在水中能防止重金属的侵入；芦荟可以治愈烧伤；山金车是消炎药；小檗消解肾结石；大蒜素有很强的杀菌作用。
化 学	合成分子性能设计稳定，即生物降解过程缓慢，能够长时间维持活性，以延长其自身寿命；植物合成化合物来发挥生物功能；植物化学品是植物体内自然生成的化合物；番木瓜能够用于伤口处理，茶树可以抵抗真菌侵袭，姜黄能改善肝功能。
物 理	孤立现象的因果关系；安慰剂效应。
工程学	废水通过沉淀、曝气和需氧消化几个过程进行处理；只有反渗透方法才能去除饮用水中几乎全部（90%）的溶解药物和来自香烟的尼古丁。
经济学	城市公共服务消耗的能源中，20%用来净化污水；医疗保健（尤其是药物）的成本正在上升，促使药房为患者提供与品牌药效果完全一样的通用药；能有效治疗人类疾病的12000种植物化合物创造了相关产业和工作岗位。
伦理学	制药业生产出昂贵的药物，但这些药物只有部分被患者消耗，却在水中保持很长时间的活性。人们明明知道药物含量随着消费增长而积累，更大剂量的化学药品没有经过任何方式去除就被释放到水体中，怎么能宣称在水中被稀释的药物没有危险？
历 史	阿育吠陀医药起源于3000年前，它包括早在中世纪就闻名于世的制剂和手术步骤；中医起源于2000多年前。
地 理	不丹通过幸福感来衡量进步。
数 学	不丹政府进行了详细的调查，来计算国民幸福指数。
生活方式	中医包括药物、针灸、推拿、气功和食疗；把将要过期的药物等废物通过厕所冲走的习惯，是一个主要但相对不引人注意的健康危机；法国人不怎么注重个人卫生，但很关注肝脏健康；美国人的保健成本是世界上最高的；持续的压力和恐惧可以改变生物系统，甚至导致各种疾病，如心脏病、中风和糖尿病；长期的愤怒和焦虑可以加速动脉粥样硬化和炎症；积极乐观的生活态度会让患冠心病的危险降低一半；连续暴露在抗生素环境下会降低免疫系统应对自然变化的能力。
社会学	医生所穿的白大褂表示效率和卫生，而医院里哗哗作响的医学机器显示了高科技的超凡能力；如果患病是因为抽烟所致，那么患者获得同情的可能性会低一些。
心理学	积极乐观的态度有助于身体恢复；个人态度创造积极或消极的身心联系；不能把某种疾病归咎于任何人，人们需要下决心挺过它。
系统论	身心一体；如果心灵不幸福，身体不会健康。

教师与家长指南

情感智慧
Emotional Intelligence

青蛙

青蛙控制着他的情绪，即便得到坏消息也不会过于压抑。他确认了水上飘所说的现象属实。青蛙对自身的局限性诚实而坦率。然而，当讨论转向医药的时候，青蛙显示出理解力，并分享他的意见，强烈表达他对人类和生物多样性遭受破坏的负面看法。青蛙向水上飘学习，渴望去了解更多，他不满足于大致的了解，提出了与个人请求相关的问题：是否有一种让人快乐的药物？

水上飘

水上飘心烦意乱，焦躁不安。他抱怨人类所犯的错误，以及在解决问题上的技术不足。水上飘脑海中有清晰的解决方案，也就是从合成药物转向天然药物，并歌颂许多文化的智慧。但水上飘也平衡了自己的意见，指出西医改善了生活质量。他的关注点从水转向更冷静达观的生活态度。他的逻辑结构有条理，不是在药物方面阐述他的观点，而是转而讨论态度和生活方式。水上飘概述了怎样管理危机，并提供在他自身行为基础上的建议，表明他有一种快乐的生活方式。

艺术
The Arts

想象一下你很生气，你会选用哪些颜色来绘画？现在，设想你感到快乐和放松，你会选什么颜色？想象你生病了，需要很长时间才能恢复，那么哪些颜色能表达你的感觉？再设想你有严重的感冒，但相信明天就会恢复，那么你现在想选用哪些颜色？颜色反映我们的心理状态、态度和期待，要用让你感到快乐和健康的颜色来绘画。哪些颜色能让你有这种感觉呢？与你的朋友们讨论你偏爱的颜色，并发现他们对颜色的选择。

TEACHER AND PARENT GUIDE

思维拓展
Systems: Making the Connections

水资源有限，被污染的水会危及生命。合成药物不易降解，被人体消化且过了有效期之后，仍然能在水中发挥药效。这看似没有意义，但它证明了现代社会的线性思维，即只考虑单一、可测变量的简单因果逻辑关系，在本例中就是用药物杀死病毒或传染性细菌。虽然某种药品在水体中可能是被稀释得浓度太低，很难被检测到，但成千上万种合成制药产品都被释放到同一水体中时，问题就出现了。随着时间推移，药物分子积累超过一个危险的阈值；即便这种积累效果还不足以直接影响人类健康，成百万的其他水生物种有可能受到冲击。海洋中的塑料制品已经积累到可以形成塑料岛屿，而鱼类可能因为错把这些塑料当作浮游生物食用，导致肠道堵塞。水中的药品还包括能从根本上改变身体和活细胞的功能与强度的激素和抗生素，这涉及从免疫系统到繁殖能力等各个方面。虽然我们这个社会总体上没有意识到由现代医学所导致的挑战，然而它也积极接受由非传统医学所提倡的健康福祉。现代医学关注因果关系，而其他医学系统关心心灵、精神、生活方式和态度。正如我们不能再在生活上忽视心理的重要性那样，我们也不能忽略由合成化学品造成的破坏。越来越多的科学证据表明我们的心理状态和态度确实影响我们的健康。我们是该学会变得快乐，并且在面对一连串负面消息时也能努力去寻找机遇。我们必须保持头脑清醒、积极和专注，同时定期找时间笑一笑，有时甚至可以自嘲。

动手能力
Capacity to Implement

努力把快乐带给不快乐的人。这样做的难度有多大？想想你比较亲近的家庭和朋友圈中那些不快乐的人。你不需要知道他（们）为什么不快乐，而是需要制定计划去振奋他（们）的精神。试着讲一个笑话，但需要注意，仅仅打扮成一个小丑是不够的，你需要做更多。

教师与家长指南

故事灵感来自

皮尔丽特·弗朗德兰
Pierrette Flandrin

皮尔丽特·弗朗德兰出生于摩洛哥的卡萨布兰卡，在这座城市长大，她父亲是宫廷的著名摄影师。她学习秘书服务专业之后，决定去巴黎生活。当她结束在巴黎银行长期从事的银行业工作并退休后，决定将她的余生投入到促进和平和理解的事业中，特别是（包括摩洛哥、阿尔及利亚和突尼斯在内的）马格里布地区和法国之间的和平与理解。她受了挫折也不气馁，特别是不被健康原因打败，总是乐观积极、平易近人、努力工作。在寻找机遇时，她保持冷静，控制好自己的情绪。她确保了《冈特生态童书》第一辑法文版的翻译工作顺利完成，对翻译细节精益求精，动员志愿者并依靠他们完成了严谨的出版目标。她在辞世之前享受到了阅读《冈特生态童书》法文版和英文版的乐趣，给我们留下了无限的回忆。

更多资讯

http://www.scielo.br/scielo.php?script=sci_arttext&pid=S0102-695X2008000400020

http://en.wikipedia.org/wiki/List_of_forms_of_alternative_medicine

http://www.hsph.harvard.edu/news/magazine/happiness-stress-heart-disease/

图书在版编目（CIP）数据

快乐而健康：汉英对照 /（比）鲍利著；唐继荣译 . －－ 上海：学林出版社，2015.6
（冈特生态童书 . 第 2 辑）
ISBN 978-7-5486-0878-3

Ⅰ . ①快… Ⅱ . ①鲍… ②唐… Ⅲ . ①生态环境－环境保护－儿童读物－汉、英 Ⅳ . ① X171.1-49

中国版本图书馆 CIP 数据核字 (2015) 第 092522 号

——————————————————————————

© 2015 Gunter Pauli
著作权合同登记号 图字 09-2015-446 号

冈特生态童书
快乐而健康

作　　者	——	冈特·鲍利
译　　者	——	唐继荣
策　　划	——	匡志强
责任编辑	——	匡志强　蔡雩奇
装帧设计	——	魏　来
出　　版	——	上海世纪出版股份有限公司 学林出版社
		地　址：上海钦州南路 81 号　　电话／传真：021-64515005
		网址：www.xuelinpress.com
发　　行	——	上海世纪出版股份有限公司发行中心
		（上海福建中路 193 号　网址：www.ewen.co）
印　　刷	——	上海图宇印刷有限公司
开　　本	——	710×1020　1/16
印　　张	——	2
字　　数	——	5 万
版　　次	——	2015 年 6 月第 1 版
		2015 年 6 月第 1 次印刷
书　　号	——	ISBN 978-7-5486-0878-3/G·327
定　　价	——	10.00 元

（如发生印刷、装订质量问题，读者可向工厂调换）